Standard candles

ALICE MAJOR

Standard candles

The University of Alberta Press

Published by

The University of Alberta Press
Ring House 2
Edmonton, Alberta, Canada
T6G 2E1
www.uap.ualberta.ca

Library and Archives Canada Cataloguing in Publication

Major, Alice, author
Standard candles / Alice Major.

(Robert Kroetsch series)
Issued in print and electronic formats.
ISBN 978-1-77212-091-2 (paperback).—
ISBN 978-1-77212-116-2 (epub).—
ISBN 978-1-77212-117-9 (kindle).—
ISBN 978-1-77212-118-6 (pdf)

I. Title.
II. Series: Robert Kroetsch series

PS8576.A515S78 2015 C811'.54
C2015-904183-X
C2015-904184-8

First edition, first printing, 2015.
Printed and bound in Canada by Houghton Boston Printers, Saskatoon, Saskatchewan.
Copyediting and proofreading by Peter Midgley.

A volume in the Robert Kroetsch Series.

The University of Alberta Press is committed to protecting our natural environment. As part of our efforts, this book is printed on Enviro Paper: it contains 100% post-consumer recycled fibres and is acid- and chlorine-free.

The University of Alberta Press gratefully acknowledges the support received for its publishing program from The Canada Council for the Arts. The University of Alberta Press also gratefully acknowledges the financial support of the Government of Canada through the Canada Book Fund (CBF) and the Government of Alberta through the Alberta Media Fund (AMF) for its publishing activities.

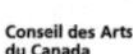

For David, always

and in memory of Monica Jean Ellis

Contents

xi Sonnet for Valentine's

The set of all gods

2 The god of prime numbers
3 The god of infinities
4 The god of symmetry
5 The god of gravity
6 The god of salt
7 The god of kites and darts
8 The god of quantum uncertainty
9 The god of probabilities
10 The baker god
11 The god of automata
12 The god of teapots
14 The god of cats
16 The god of sparrows
17 The god of hearts
18 The jeweller god
19 The god of dark
20 The god of memory
21 The muse of universes

Ordinary matter

24 Ordinary matter
26 Vacuum fluctuations
27 The helium thoughts

28 Advice to the lovelorn
30 Three-body problem
31 Love in three dimensions
32 Heavy elements
33 Local bubble
34 Catechism

Standard candles

40 1 Address | 1959
42 2 Clouds of glory | 1908
45 3 Pythagorean theorem | 1965
46 4 Triangulation | 1808
49 5 The end of greatness | 2000
51 6 In the Castle of Stars | 1576
54 7 Supernova Type 1 A | 1997
57 8 Then death returns
59 9 In all that void
60 10 Looking out to the dark | December 1928
62 11 $d = (X-x)^2 + (Y-y)^2 + (Z-z)^2 - c(T-t)^2$ | now
64 A prayer to bring you home

Muscle of difficulty

68 Muscle of difficulty
70 Yet another crack in the foundation
72 Day's eye
74 Expanding space
76 The movers' dilemma
78 Rectangularization of the morbidity curve
80 Now, that amphibious moment
81 To the generations that will live a thousand years
82 Last scattering surface

Let us compare cosmologies

86 1 Let us compare cosmologies
87 2 The Orphic follower
88 3 A pope
89 4 The evangelist
90 5 The philosophical skeptic
91 6 The nihilist
92 7 The totalitarian
93 8 The survivalist
94 9 The optimist
95 10 The magician
96 11 The baker
97 12 The consumer
98 13 The funeral director
99 14 The Manichean
100 15 The artist

Sins and virtues

102 Avarice
104 Lust
105 Gluttony
107 Envy
108 Pride
110 Anger
111 Sloth
113 Mercy
115 Hope

Shifting wavelengths

118 Tortoise and fern
120 Fingers of God

121 The barber's paradox
122 Zeno's paradox
123 Twin paradox
124 Sand reckonings: Eubulides' paradox
125 Honeycomb conjectures
127 Bee violet
128 Optical molasses
129 Life adapts to inhospitable environments
130 How to tell a Martian my heart is on the left

Underworlds

132 Persephone and I are underground
135 The outer dark
138 Cocytus
142 Niflheim
145 The man with no hands
147 Each of us the centre of a circle

Postscript

150 God submits a grant application to the Canada Council

153 Notes
163 Acknowledgements

Sonnet for Valentine's *for David*

Go, little poem, into the space between
planets, across the unbounded page
inscribed by stars. A tiny, ticking machine
of levers and polished surfaces—
clear evidence of intent, design.
Let the aliens who intercept it
learn the virtues of this love of mine,
his kindly constellation. Let them share
my wonder at the dense relationship
of soul to smile within the dear,
dear boundaries of skin. Go little ship
of space beyond the gravity of time,
        and, beating always, prove
        there is indeed a god
                        of love.

We keep searching for the one creating deity, the theory of everything, the god particle, the basic equation to drive our unfolding universe. Surely, we think, we can isolate one single, monotheistic grain from which everything else is built.

What if, instead, our world is built by an interlocking pantheon, a set of gods? All of them (each of them) central, essential, supreme...

The set of all gods

The god of prime numbers

—trinity, quintic, indivisible seven—
visits her creation
often in its early moments

then draws away for ever-lengthening periods

oh, how long must we inhabit
a dreary world of common factors
'til her return?

The god of infinities

is wizened, smaller than the space
between "one-over-n"
and one   gets tinier and tinier
world without end
and then

The god of symmetry

says *fiat lux*

not with a mighty groan of light
but in a whisper
that blows the smallest crease
            cramp crimp
into perfect equipoise

allows himself to break
            a fissure king

The god of gravity

is weak and distant   somewhere
out there

adding up masses

delicate crush

The god of salt

creates everything
in her own image

tiny crystal   shaped
like still-tinier molecule

no apparent boundaries divide
creator from created form

and shape's hegemony expands—
great clear cubes accreting

The god of kites and darts

launches into the air
flips back and forth between
good and evil   tiles
the forking universe with
contradictions snug up
against each other
sharing edges

The god of quantum uncertainty

trickster here there nowhere

immoral immortal coyote

The god of probabilities

drags up mountains
of improbability
with sharp crags
and granite sides

thereby creating the likely valleys
where we can cluster   comorbid

below the peaks where only she
may abide

The baker god

Who knows what shape we're in?
Flat cookie, doughnut-torus,
or perhaps the crazy twistings of a cruller.

The baker god, kneader into shapes,
come down to earth and sitting
in the dunkin' donut shop. Methodically
he tries them all—the two-holed torus,
the simple solid ones,
the folded-over-sealed ones with lemon filling
(how does it get inside?)

The universes sit in trays
with party-coloured sprinkles on their sugared tops.
The baker god turns them out repeatedly
in batches.

The god of automata

links atom to atom
like a knitting pattern
with simple rules
—*when a, respond with b*—
and algal blooms rise and fall,
vast populations of the stupid.

The god of automata
crochets a chain
of mindless proteins
into a loop. She winds a cord,
flicks messages along the fibres.
Muscles twitch as axon fires
the dim bulb of neuron.

Against time's cycles,
she struggles
to make her frail creations
coherent, urges them
to unite into wisdom

before their nets collapse
under their own dumb weight.

The god of teapots

You are corpulent and unworried.
You accept what pours in
and pour it out—

amber, tan, sepia,
the percolations of brown,
the brewed colour of peat,
muskeg, spruce bog,
wetlands.

You retain traces
of vegetable digestion.
A crust of memory lines you—
a biofilm, a plaque
that flavours the ongoing.

Topologically speaking, you are
a two-holed torus.
Plain clay stretched, scooped,
spouted, handled.

Heat has come and gone
in your history. You take it in
and let it radiate away.

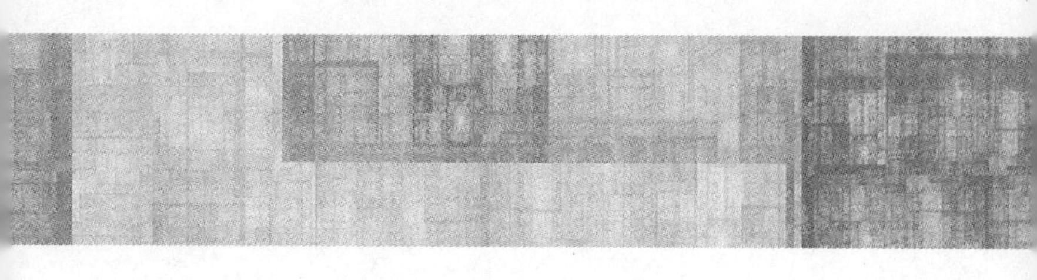

Your shape imprinted
in the hard heat of firing,
which you remember to this day.

The god of cats

World-Cat
uncoiled her tail and leapt
on the flecked back of Sky Antelope,
pulled it down in the death shriek
that began to hunt
time down.
Cat devoured
a space for us, her soft-pawed
descendants, in the belly of
her prey.
Its meaty haunch became
the earth on which we prowl.
Its flesh hatches into mice and voles
and small sweet birds.
Its rib cage
holds up the sky. At night,
we see a single arch of bone,
a white span of heaven.
World-Cat
slid her inner eyelid closed
and hid invisible behind it.
But still
sustains all being with her purr.

That deep and rumbling rhythm underlies
the rise and fall of birds in flight,
the interplay of hunger
and plenty.
We stretch
close to her heartbeat as we can
and repeat her mythic breathing—a tribute,
our ecstatic contribution
to holding up
the world.

The god of sparrows

observes
the multifarious detail,
the minutiae of feather, fact,
flight, flick, fleck, egg,
          mite

is
a bright round eye

The god of hearts

Indra the drummer,
to whom all things bow down,
bangs on and on
behind the heavy-metal band
Jupiter Tonans. Sweat
spangles his bandana,
his bulging muscles pulse
tattoos. Over the hammer
he lays a throb of high-hat,
strokes a shimmer
onto cymbals, scats
staccato patterns.
The crowd swaying in the pit
goes wild with worship,
caught up in the clanging heat.
Don't stop
don't stop don't
stop don't stop

But it's the tattooed drummer
who measures out the beat

and it ends.

The jeweller god

Out there, off-centre in Centauri,
a dim diamond sun is set
in the deep kimberlite of space.
Its core of transmuting carbon
crystallizes slowly, carat by carat,
under gravity's contracting purgatory.

Other stars end grimly in an iron limbo
or collapse to a nothingness so intense
they pull the black cloth of space-time
into a distorted shroud around them.

But a few, the rare elect,
shrink to this white-dwarf twinkle—
perfected for the god's ring finger.

The god of dark

muscles himself into the universe
through its keyhole

a menacing energy
that shoves us apart
imposes the sentence of tense—

future isolation
locked past

The god of memory

is a telescope, pointed at the past
and patiently collecting
every fragmentary package
of fading wavelengths
from the universal mesh of ripples.

Wait. Don't worry.
It's all here, somewhere.
Everything.

The muse of universes

Once in a trillion years
the muse of universes
claps her hands. And, with that shock
of light, reverses

an aeon of drift, dilution,
the outward-rolling wave
of dark and the illusion
of end times.

A new draft, she orders
and the universe erupts
into rhyme, fields and forces
echoing. She rebuts

formlessness, sparks stanzas
from an alphabet of particles,
spells out what matters, what
radiates, what tickles

the fancy into galaxies
with gravity's feather pen.
She unrolls the scroll of space,
says, *There. Now try again.*

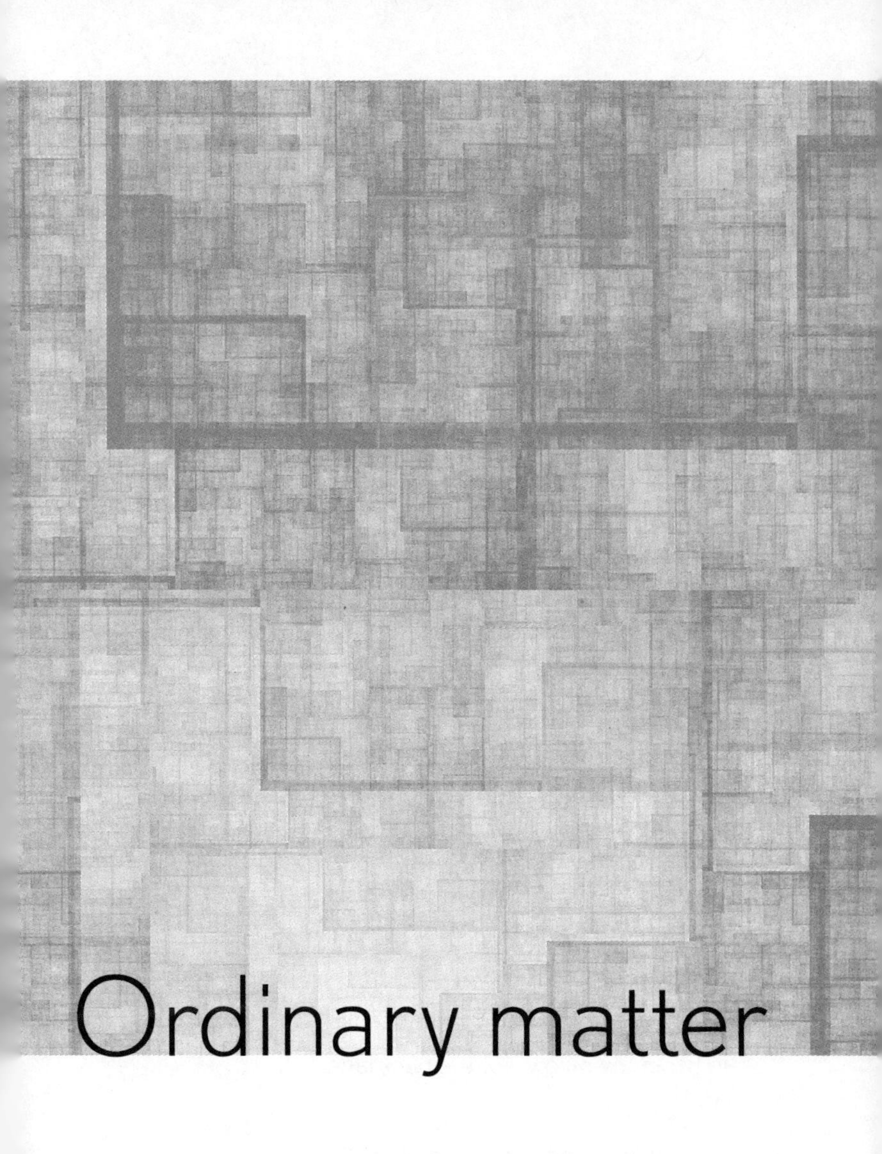

Ordinary matter

Ordinary matter

I am yanking at clematis,
pulling down the strands
that have tangled into a tower
of summer. Rapunzel's tresses
in reverse, hiding windows,
billowing over into eavestroughs.

Sails of leaves spread
to catch sunlight, silky seedheads
flare to catch at air. The sheer
delight of taking up space.

I stomp the stems into compression,
packages that can be fed into the maw
of orange plastic bags—
that whole golden bower reduced
to two puffy pillows for the back lane.

And I think of how much smaller
they could be, since matter
is composed of so much empty
space between spinning nucleus
and distant echoing electrons.

Ordinary matter
makes up so little of the universe—
swathes and filaments of plasma,
a skiff of interstellar dust

And yet these few percentage points
of substance are determined to erupt
into stars and clouds of gas and founts
of coalescing molecules.

The window hidden by the vine
has now been cleared. My heart shouts
Jump, Rapunzel!
Jump into the arms of matter—trust
to its enormous bounce.

Vacuum fluctuations

Science speculates the universe was born
from a random vacuum fluctuation.

Looking round my living room, it could be true.

The place is silting up with spring's loess—
shedding cat hair, sand tracked in
from ice-covered paths, potting soil
from the plant upended by the kitten, and yes

there sure could be a baby universe in there
rolling around below the sideboard
like a cat toy with a tinkle at its centre.

I should get the vacuum out, get back
to original conditions, smooth out the quantum foam.

Or maybe leave it all alone, continue writing poems
randomly, see what inflates next
to bring a new world into being—
my littered pre-text.

The helium thoughts

Stars rotate steadily on their axes
because each always thinks the same thoughts
about the same things.
Plato, Timaeus (40A)

Our starry brains—their frail shells stuffed
with as many neurons as the galaxy has suns.
On this sunlit afternoon, I clutch
my temples to keep the giant number in—
more synaptic links than there are seconds
in thirty million years. The combinatorics
make me spin. We could think anything.

And yet the helium thoughts of youth consume us.
The paths of thought are bound in myelin,
spiral arms that wrap us tightly in the past—our own
and the bonds of evolution. Thought's locks spin
in pre-set combinations and conclusions.
It's hard to get past helium
when we are still so young.

Advice to the lovelorn

We are observing Eros,
raddled asteroid, lumpish and erratic,
on a loopy path such that
our planet and that planetoid
just might find themselves
one day trying to inhabit
the same point on their intersecting lines.

So keep a wary eye on it.
It's no sleek bow-directed arrow,
this potato-shaped tumbler
aimed at no particular target in the dark—
just a stranger across a room
that could suddenly become
much too crowded.

Remember this and contemplate
the drowned crater of Chicxulub,
where an asteroid stove in the planet's rib,
turned rock to instant liquid
and instantly back to mountain, a crumpled scab.
And how the noise was heard two thousand miles away—
everyone knew your business. How dark
the skies turned, whole species lying down
to die in the shadows.

But remember too—
you can't see that crater now.
The blue gulf washes over it.
Small creatures you'd previously ignored
became more interesting, filled in the gaps.

And always remember,
though nothing will ever be the same,
still you are bigger
than it is.

Three-body problem

Two bodies present no
insoluble equations.
Their motions elegant, simple,
they will orbit each other forever
tracing a perfect ellipse
around the shared heart.

But add the smallest satellite—
weight of an unborn child, lovers
in the most transitory
of conjunctions,
the weight of a memory
circling in the dark—

and no one can calculate
when the instabilities begin,
when ellipse may evolve
into spiral, into tangent,
into the geometry
of heartbreak.

Love in three dimensions

Plastic wrap hugs smoothly to its roll
around a cardboard heart
'til I apply sufficient force to pull
a piece of clinging film apart

and the edge scallops into a fractal froufrou,
cascade of ripplets, larger waves.
This, from the same fantastic algorithms
that crinkle flower petals,
lettuce leaves—buckled surfaces
that refuse flatness,
cannot be ironed out.

If such edges lived in four dimensions,
their pattern would persist
as simple curves, serene expansions,
their convoluted arcs relaxed.

But in our three-dimensional experience, tensions
are forced to find their way to ripples
at the edge
and in this severed space, we must learn to love
our ornate, wrinkled shores
and crimped perimeters.

Heavy elements

First loves come lightly, fused
from the simplest elements—dizzy helium
compressed from casual couplings
in the first swirl of stars.

But those first, bright burnings
cannot build planets. They eat
their hearts out and collapse
to leave a taste of iron.

From their heavy-hearted remnants
the new stars coalesce, pull planets
into orbit, endow them with granite
and gold.

Local bubble

The Local Bubble. A calmed space
swept clear by trauma—the aftermath
of old, exploded stars.

This bubble-room. Your calm face
across its space, absorbed in the sweep
of your gentle fingers scrolling on a screen.

This wooden table-top is screened by lace—
threads looped and twisted to make openings,
patterned flowers, figures against ground.

Beyond the window, April's winds lambaste
branches gnarled with buds, ready to explode
into the elements of leaf and sepal.

A bubbling world, embraced
by the galaxy's scrolling arms, the calm
after chaos that opens space.

Catechism

Four questions for winter

What does the roof know?

The tickle and scritch of bird feet.
The aching angle
of an arm bent to protect.

Who is the snow?

Moon's admirer and echo,
friend of white owls.

Where does the world end?

Over the back hedge.
In the sigh of a stopped furnace
under the stairs.

Where do I love you?

Here.

Three questions for spring

Why is the sky blue?

It's a colour that washes well,
hung out to dry on a line
of white gulls.

Where do babies come from?

They are ordered from the catalogue
of the universe wishing.

Why do I love you?

You are blue.
You wash well.
You are from my catalogue
of answered wishes.

Three questions for summer

Where does the leaf hide?

In full view. In profusion,
in fused green.

What does the mosquito seek?

The obvious. Glut.
The unspeakable. Veins under cover.
Pumped capillaries.

What does thunder love?

The sound of its own voice
shouting at the corridors of cloud
I love you. I love you.

Four questions for autumn

If not rain, who?

The loud wind pulling down
dry leaves. They click pavement
like drops,
shower and shower and shower.

Why am I here?

To tuck the wasps in.
Their feeble paper houses
will tatter with winter.

When does the window close?

When the geese leave
honking V V V
for vast, for vista, for variorum.

What is the candle for?

Emergencies of love, the flare
that whispers here
over here, follow me
in falling dusk.

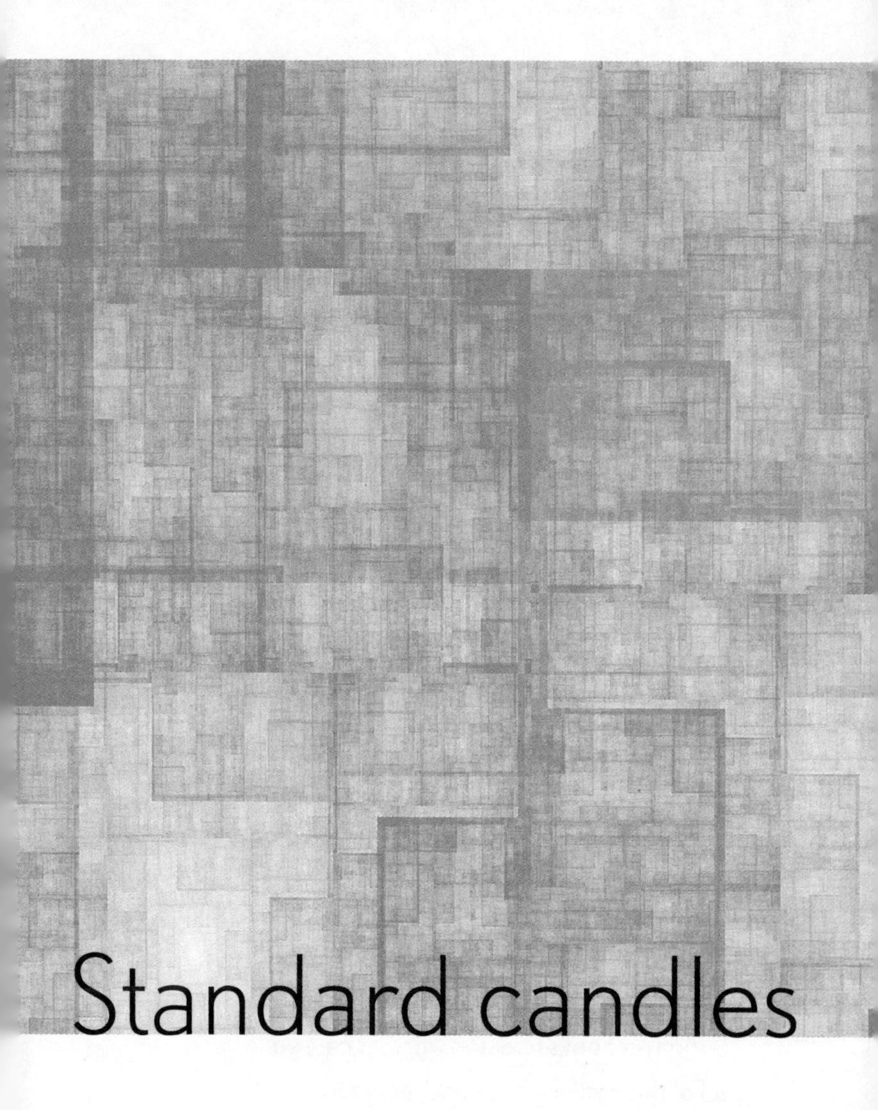

Standard candles

1 **Address** 1959

Pell Street. Scarborough, Ontario,
she writes in a child's round hand, to go
onward and outward: *Canada, The Earth,*
The Solar System, Milky Way. The Universe.

Delighted at how everything spreads out from her.
Ten-year-old astronomer
learning the track Orion takes across
the crisp gasp of winter sky. She is lost
in the marvellous stars—
this universe like a set of nested spheres
with her family's basement living room
like a beige-walled pericardium
wrapped around its heart.

And somewhere out there, there is God.
Sunday school's Creator, who has caused
all of this vast envelope addressed to her.
God who is the measurer
and measure of all that hierarchy
of distances, out to the dark so sparkly
with patterned stars.

But there was somewhere else. A place
we'd never reach with rockets nosing space.
Heaven, where Nana and her aunt had gone
and where the dead went living on.
Another place, apart.

2 Clouds of glory 1908

Miss Leavitt, lace-waisted,
hair knotted neatly
at her neck's nape, pores
over photographic plates.
In the observatory's
panelled, gas-lit room,
Henrietta exercises
her gift for the meticulous.

These slabs of glass have caught
the light of constellations
from the high skies of the Andes—
the Magellanic Clouds,
those flags of glowing gauze
that guided mariners across
the unknown southern oceans.

Magellan, Tasman, Cook
and, long before them,
those ancient wanderers
whose canoes and skilful rafts
reached archipelagos
and islands and scattered
necklaces of coral.

The telescope's glass eye
resolves those clouds of glory
into a starry net—
an *island universe*, adrift
beyond our own Milky Way.
And there, as here, some stars
flash out like sweeping beams
of lighthouse beacons,
now bright, then dim.

Miss Leavitt goes exploring
with her tools of navigation
—her blink comparator,
mathematics, graphs—
to demonstrate a simple,
stunning truth: *The brighter
it is, the slower it blinks.*

She is holding up a standard candle.

Now telescopes can search
still-smaller, fainter clouds
that smudge the fathomless,
unmapped heavens.

And on those further islands find
still other lighthouse stars
to calculate how far
away they are.

For all of us to realize,
amazed, how very far
from us those islands lie.

3 Pythagorean theorem 1965

The girl learns Euclid and Pythagoras—
that theorem, luminous and ancient
as Egypt. It seems mysterious
    and beautiful, this simple shape she draws
    on a clean page with her ruler. Just
    a triangle. But obedient to laws
that bind its measurements—the length of sides
linked intimately, but not by inches notched
along line's thin dimension. They're unified
    in another space entirely—they share
    area, square. The sides are the shadow edges
    of another shape. It's like a prayer
that she is learning, this theorem
of trinity and distance—a formula
sent down to her from time, an orison
    that she can hear and learn,
    repeat within the congregation
    of the classroom, and then send on again.

4 Triangulation 1808

William Lambton climbs the tangled statuary
of Kumbeswarar—a soaring wedding cake of layered
earth and heaven, piled up with brightly painted gods.

He gazes from the temple's tallest gopuram
across the waving palm-frond floor of Tamil Nadu,
seeking another temple spire to serve

as the next dot on his network of geodesy.
The Great Arc Series, an odyssey of trigonometry
that will be drawn through jungle, over rivers all the length

of India, lugging The Great Theodolite
from Cape Comorin to the brutal terai of Nepal
and the Himalayas' massive planted triangles.

An act of empire, yes. The heavy stamp of license
on a forced marriage. Britain needs to know precisely
the dimensions of this new dark spouse.

But Lambton's quest is more. He is fired by desire
to know—to measure nothing less than earth herself,
the detailed gradations of her curvature.

To measure a universe, you start by measuring home.
To measure home, you start with a triangle.
To measure a triangle, you start with a single line

calculated with fanatical exactitude. A folding chain
of rods machined from blister steel, nested in wooden frames.
Thermometers to monitor how heat deflects length.

With that first measurement made certain, you find
a point on the horizon, swing your brass theodolite
towards it, calculate angles, points, distances

to lay an imaginary net, a mesh of measurement
founded on that first known length. Then unfold your rods again
to measure the final triangle's final side.

A miracle. If the calculated length of your last line
matches its real, rod-confirmed distance, then you know,
god-like, all your other measurements are true.

The gods of Tamil Nadu acquiesce in serving
as reference points for this ménage à trois that binds
money, politics and knowledge in a tight theology.

Though at the temple of Tanjore, Shiva shrugs,
a rope breaks, the theodolite comes bouncing to the floor.
Still, men will mend such accidents and carry on.

Lambton, pleasant Englishman, retreats to Bangalore,
patiently repairs it, sets out again. His half-caste children,
his country wife, Kummerboo, his loyal corps

of young surveyors (none *pukka sahib*), the complex skein
of conquest and kindness all come with him
in the search for that which we may know for sure.

5 The end of greatness 2000

The woman's life arcs across
decades of discovery.
Sometimes, she looks up
from triangulations of the ordinary
(work, household, words) to notice

the universe becoming bigger.
New standard candles lit—supernovae
shedding light in torrents,
beacons sweeping out and over
space, to fill the eyes of instruments

that grow ever more enormous,
complex. And yet more sensitive.
For it is from the smallest things
we measure—a curve
of luminosity, a scatterling.

Now she could amplify the address
on her envelope, add *Orion Arm,*
The Local Group (that reef of galaxies
around us—Andromeda,
Triangulum, the Magellanic Clouds),

then on again—the *Virgo Supercluster,*
The Sloan Great Wall.
Out to the limitless end of greatness—
a foam of filaments and bubbles,
the spray of centreless light.

Where is God's place in this?
she wonders, sometimes.
How can anything so vast
care for her, send messages
to her insignificant address?

But the thought is fleeting, scatters
like a photon's flashing passage
through a telescope, down the corridor
of angled mirrors
to register on some stored image—

something to be looked at later.
For now, she lets heaven go,
allows her childhood dead to scatter
into the past. Out there, somewhere.
For now, she is content to be amazed.

6 In the Castle of Stars 1576

Stjerneborg, newest of observatories.
It holds no telescopes—such ground-glass curves
wait in the future. But still the human eye
can see stars strewn across the sky
and needs careful instruments
to aid in noting their eternal patterns.

Tycho Brahe mounts his quadrants,
his astrolabes and armillary spheres,
the alidades with slotted pinnules
(his own invention) to sight
the faintest spark of light correctly,
without the winking parallax
of two eyes shifting left and right.

His instruments are ringed in wide brass arcs
etched with his clever patterns
that translate slender differences
of distance on the dome of dark
into measurements an eye can read.
Night after night, season after season,
he will note, observe, reason.

For there was a star that changed.
Stella nova. Four short years ago,
a sudden brilliant cushion, stuffed below
Cassiopeia's throne, outshining
that whole familiar constellation.

Tycho could not at first believe his eyes.
Do you see that too? he asked and asked again.
But it was there—a brilliance to be measured.
Over months, a year, he noted its position
in the constellation's loose embrace,
saw its stillness while the planets went
their wandering way around it.
He watched its colour blaze
then fade to lead. He watched
until it disappeared.

And with it, heaven had changed.
Aristotle's perfect, changeless spheres
lay crashed in littered shards.
Simply because something had been seen.

Many close their eyes, insist
on holding up cracked glass
for longer. They say it is a comet,
or a rippling in the lower atmosphere.

But Tycho thumbs his celebrated metal nose
at such slow wits and dullards,
goes on to build his castle of astronomy,
his catalogue of heaven, his great brass globe—
a celestial map engraved
with a thousand constellations.

He does, however, hold to earth—the centre
of everything. It is too slow and lazy
to be whirling round the sun, he thinks.
According to his computations, stars
could lie just past Saturn, be reasonable
in size. Planets could whirl
around the sun, while it rolled over us.

He is wrong of course, unwilling to forfeit
both common sense and scripture.
His Ptolemaic rods, triquetrum, theories
are ancient instruments inherited, refined.
They swing their smooth hinge
between the past and a vast expanding future
still unthinkable. But he is prepared to see
that heaven can change.

7 Supernova Type 1A 1997

In a universe so vast
that rare things happen often,
these star explosions sparkle constantly
in galaxies so far away their light is stretched
to a low red thrum. And yet,
the supernova's signature is unmistakable
and uniform—a silver flare of nickel
forming in the blast
as elements collapse.

Another standard candle
to be measured. Meaning to be sifted
from plotted light curves, red shifts.

This is what he's good at. *Bump hunting*.
he calls it. Spotting data clustered
like a church spire that juts above
the trees and houses hunched around it.
As if to signal, *here. There's something*
here.

The team is huddled round
Gerson's square-ruled page of graph paper.
Almost old-fashioned, it looks,
this hand-drawn histogram. Some lines painted out
with white correcting fluid, then re-inscribed.

Scribbled notes and title in an upright hand.
And that tall spire
where no spire should be.

Gerson Goldhaber, no astronomer,
hasn't seen this pattern by peering
through a telescope. He sees it in the scatter
not of light, but number,
the same way he discovered
the fingerprints of unknown particles
in data sprayed out by accelerators.
There, he made the smallest things
visible. Now he shows the far end of the scale.

That pillar on his graph:
supernovae, whose light has shifted
to impossible lengths, moving too fast
for belief. Not just one or two
that could be accidents of observation.
Too many of them piling up.

Something is happening here—a universe
blowing itself apart, space accelerating.
Those distant candles seem to gutter
on the edge of invisibility. Soon their light
will disappear past any hope of reaching us.

It can't be true, they think. His team
gathers round the graph, doubting
its cryptic implications. They will check
and check again. Gerson says
he'll keep working on it. And, rueful,
I've been known to make mistakes.

But data keeps streaming in.
His intuition holds. This is some kind of truth
to be explained. The heavens
have changed again.

8 Then death returns

The woman flies towards it, pulled by urgent
messages, emergency, collapse.
How long it takes to cross a continent.
(A day, a century.)

Beyond the angled wings, evening sky
fades to the old blue of airmail envelopes—
those tissue sheets on which you penned your scraps
of news and love, sealed down the flaps,
and turned the paper over to inscribe
an address far across an ocean.

Now she is being airmailed home
to kneel beside her mother's bed,
to stroke her head, to strive for calm
through this wrenched, red-shifted time.
To whisper, *You are tucked into my heart*,
and hear the thread of voice respond
And you are tucked in mine
always. Always.

The woman is left with ashes and no maps.
Only the bitter cry, *Where are you now*
to hold me in your heart? Anguish stabs
its hole in her seamless world,

rips out one essential point, explodes
the punctured topology of globe
into a flayed plane stretched to the edge
of everything. An incommensurable page—
no co-ordinates, no spires or temples,
no marks or margins, no instrument, no theorem
by which to understand its distances
and tie them into something relative
to her. A universe of loss. Within its gaps
she must find a place to live.

9 In all that void

Where are you, oh my dead?
The night is icy and immense.
Winter's stars are rising.
I cannot be comforted.

I cannot be comforted
by childhood's fantasies
of other worlds in parallel,
of heavens where the dead

live loving on, not dead.
No candles light me to the place
where I might hold you in my arms again.
The space without you spreads

to infinity. In all that void, no shred
of you to love me,
no space for those realms
where I might meet again the dead.

The night is icy and immense.
I can not be comforted.

10 Looking out to the dark December 1928

Miss Leavitt, measuring the magnitude of stars.
Your grave weeps with funeral wreaths,
petals frozen to pale transparency.

The mourners have now gone. There will be
only your name etched on a granite stone
in block letters, along with the small bones
of a baby brother, baby sister, you hardly knew.

There will be your tiny, careful lettering
in the ledger where you noted stars—
their exact positions, minute gradations
of light. Later, there will be a lunar crater
named for you. But though you had
"a nature full of sunshine" we will choose
a bashed basin on the dark side of the moon,
facing always away, towards oblivion.

This is all we now retain of you—
a name, a few scraps of obituary,
an early death, the numbers on the blue-
ruled pages that you recorded faithfully.

Faithful. Henrietta, daughter and descendant
of the clergy, "sincere in your attachment
to religion and your church," did it worry you?

This great lurch outward, your discovery
we are a single blinking island
on a swelling sea? Were you lost
on that dark ocean? Or no.
You loved your glowing clouds.

11 $d = (X-x)^2 + (Y-y)^2 + (Z-z)^2 - c(T-t)^2$ now

Everything in the universe is held
(*everything*. The whole bewildering totality
of star and atom, event and void)

in a net of connection, the overlapping sea
of space-time
where every distance (through time, through space) can be

measured from that slant line,
hypotenuse, by the old geometrist's
theory of triangle. Events combine

in patterns that may stretch and shift
yet stay forever linked—a faith beneath
all partings, on which geometry insists.

She has walked the baseline of grief,
step by step, wondering what point
to fix on in the circling funeral wreath

of galaxies. There is no point
where the dead can wait for her unchanged.
No future heaven out of time. Only the faint

and fading ripples of the past, retained
as best she can. Remembered touch
of hands that rearranged

her hair and pointed to Orion. This much
there will always be. This evidence
of closeness, relativity, to etch

upon the fine glass instruments
of the heart. She turns to the tiny distances
of home as points of reference—

spacing of photos on the mantelpieces,
brushes and comb upon a dresser.
From such angles, cosines, spaces

she will slowly try to measure
out to the stars. From this home address
in the realm of starry, vast forever

she paces out the length of earthliness—
the final, triangulated measurement
that verifies the rest.

A prayer to bring you home

Come home.
The street is lined with green ash.
You know these trees, now turning bronze.
Come home.
You know the songs
sung in the cracked voice of this sidewalk.
Come home
past the drying stalks
of morning, mourning, sigh and clutter.
Come home
through the litter
of autumn leaving us again.

Come home.
I am watching for you
from the window, half empty glass.
Come home
up the path
you have always known.
Come home.
Your suitcase is heavy as a headstone,
light as a purseful of leaves.

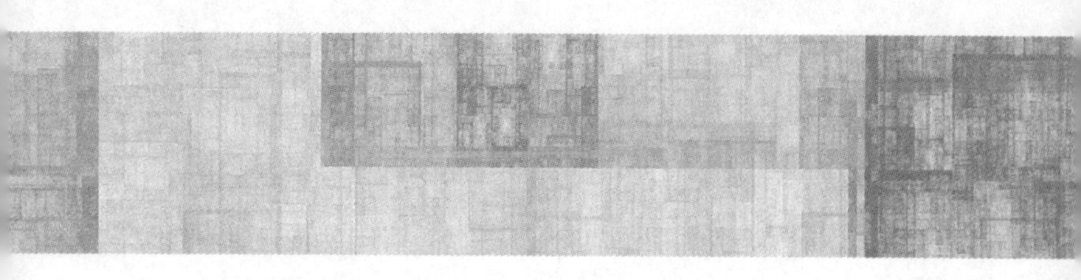

Come home.
It is warm.
Come in
my arms.

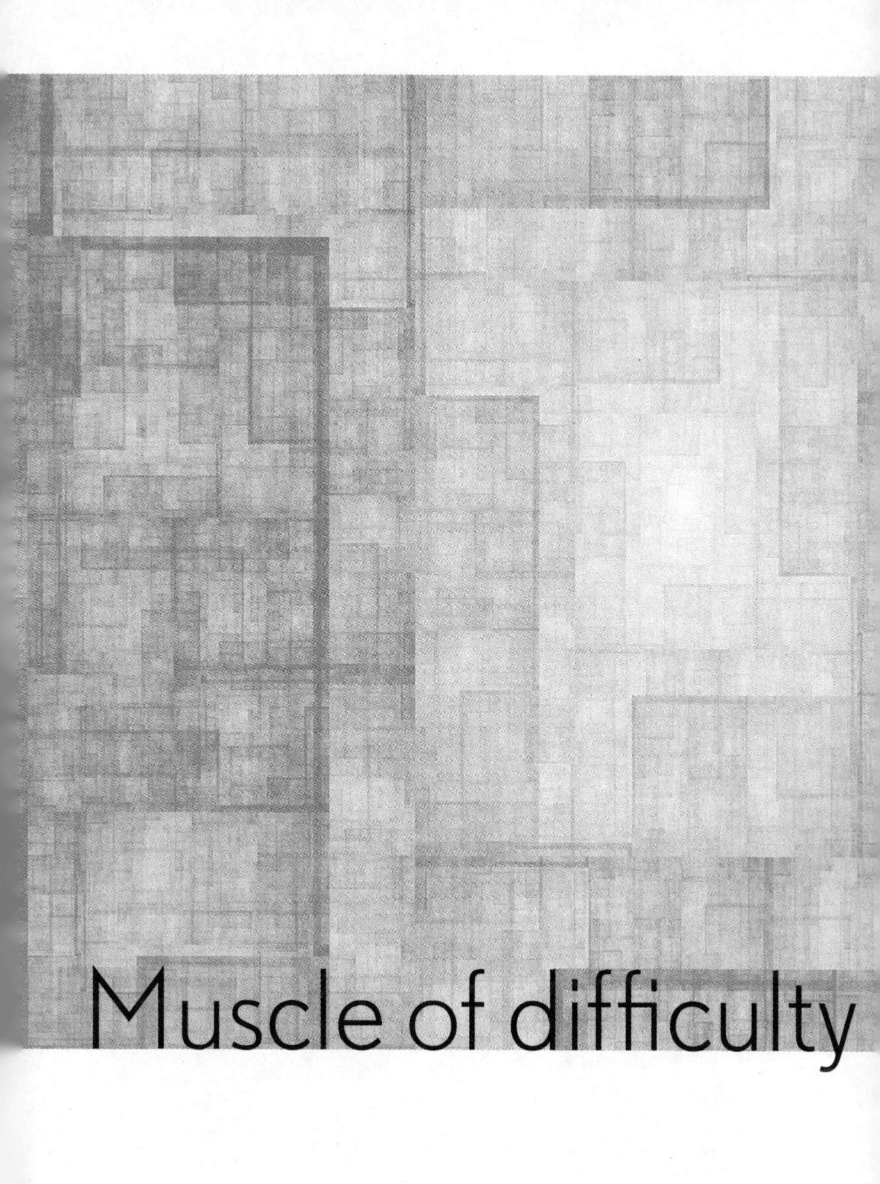

Muscle of difficulty

Muscle of difficulty

November's clench. A sullen band
of cloud is louring in the West—
a low forehead, a corrugated frown.
Behind it comes the cold drop of frost
and autumn's first hard night.

Corrugator—the tightening band
over the forehead's bone,
the "muscle of difficulty,"
of concentration, effort, of leaning in
to frigid wind.

I lean into this coming season
of difficulty, when the sun
will struggle to raise its head
above the angle of sunset,
its bleak obliquity.

November's forehead wears
scoured furrow, tension.
Forgets joy, the orbicular crinkle
of eye, the other muscles
to be strengthened.

I think of squinting into the ache
of snow, corrugated tracks.
Facing into November, I find it
difficult to anticipate
consolations—

the warmth of small, enclosed spaces,
the candles of memory
at its centre. How can this ever be
enough? I fear too much.
The losses. Isolation.

Yet another crack in the foundation

has opened its black spine
in the pale cement of the basement wall.
Not a new line—
just an old complaint that bends and flexes.

This elderly dwelling
with its porches, footless extensions
and chitinous plating
creeps in the surrounding clay

like an arthropod
swinging its segmented exoskeleton.
Its arthritic plod
records the seasons' thaws and freezings.

We must patch joints
in the armour's rusting boundary.
Vulnerable points
keep opening at knee, neck, groin.

And thus repair guys
must be called in. We do not wish to join
the moles and pismires.
Not yet. Outside must stay out

and walls kept sound to turn
the points of water, the spider's tiny daggers,
and keep the worm's
soft spear from piercing this stone cuirass.

We must preserve
this sarcophagus already half submerged
in earth.

Day's eye

Today's August intensity—
a blaze of breeze-tossed sun,
multidimensional light, reckless
orange of zinnia, my gay
sunflowers, hemerocallis
(*beautiful for a day*).

At its heart, a density—
a microscopic black hole.
A small mew quieted, so small
not even the black ants on the path
can hear its loss.
A little cat, last seen wrapped
in a blue blanket,
his eye suddenly vacant
as though we folded a florist's paper cone
around an armful of daisies.

So tiny an immensity
floating in the day's gold air
near eye level, wherever
I gaze. Like the spot that hovers
identical in the centre
of every streetlight after nightfall.

So small. But a part
of growing darkness at the heart
of our galaxy—the vast, vacant eye
that centres our flowering spray
of stars, relentless.

Expanding space

Straight, white jet trail
holds the two halves of heaven
together, like a zipper.

Space is growing bigger all the time,
as the empty regions between galaxies
swell, push apart the fine detail,
the sparkling buttons.

My father and the little dog admire,
one—the blue cloth of sky,
one—the fall-snuffled earth
with its park litter, its low-slung emulsion
of jackrabbit, common-or-garden cat,
chicken bones from dumpsters
raided by ravens.

We take the path up and over
a small hill, a man-made swelling
planted with saplings. My father points
to the blue, tells me yet again of moors
climbed with his father. One memory
of the few that linger still for him, a trace
of long-past passage.

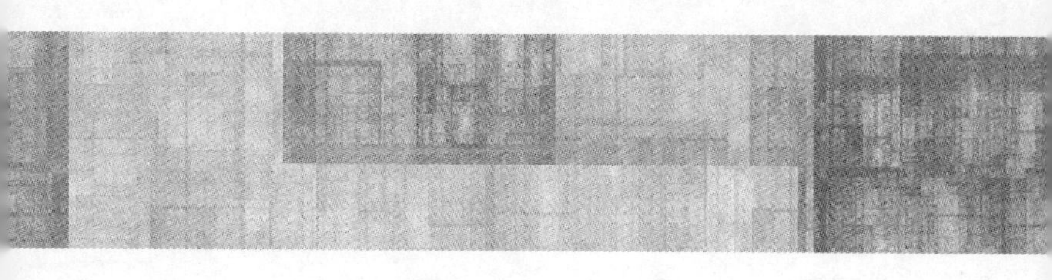

No other detail to that vast interior
of sky. Just the white seam,
its zippered, interlacing coils
spreading at the centre.

The movers' dilemma

There is a puzzle. It includes
words like *Pythagoras*, *hypotenuse*.

There is a hallway, papered in pale ferns.
It comes to a right angle, a hard turn.

Back there, an apartment's open door reveals
birdsong through windows, elegant pastels.

Ahead, around the corner, wait the sliding arms
of a metal elevator. Silent. The shaft stands

ready to descend to depths below,
territory that our hearts dread to know.

At the hall's sharp angle, sticks
a forced hypotenuse, a sofa. The trick

is how to move it further. On its length
a girl is curled, saving her fading strength.

The movers' dilemma. We cannot return
to the bright door behind us, shedding sun.

Nor do we wish to push the bier ahead,
around the corner, to the corridor's feared end.

And so we wait, make tea, light candles,
laugh together in this *pro tempore* triangle.

The calculations of Pythagoras enclose our fortune.
If the walls were wider, or the sofa shortened,

then the corner would be turned. As it must be
someday, through pain's implacable geometry.

The girl will shrink a little. With that space let in,
the walls will soften like a womb, accepting.

And we must pick up grief on aching shoulders
and move to where the other arms enfold her.

Rectangularization of the morbidity curve

The demographer's desired geometry:
a hale old age and then a rapid fall—
a sudden tumble down a slope of scree
to that inevitable, oblong box.

And we agree.
"Oh, that's the way I want to go."
The unexpected failure of an artery
from which we never waken. We dread

the lingering morbidities ahead,
slow occlusions of cognition, bodies
angled into immobility, diapered,
and life a sad, sagged line, tethered

to an all-too-distant other end.
But then I think of you. You said it too:
"Just put me on an ice floe, send
me off. I won't want to live."

And yet you did. At no point on that curve
of long diminishment
were you prepared to leave,
to face the cliff. You wanted life until its end.

Then I think of curves—the gull's wing drawn
from lifting shoulder to the tapered tip
trailing its final feather into air. A line
lovelier, perhaps, than that sharp edge

of rock plunging to ocean. How to end
this? There will be the narrow plot,
the dug rectangle. Until then
there will be seagulls wheeling over headland.

Now, that amphibious moment

between past and future.
Zero. Neither negative nor positive,
the narrowest of no-man's lands
between two kingdoms.

And I cannot share it with you any longer.
We have the past together. Your life
tangled in my memory.
I carry you with me into the future.
In whatever kingdoms I will travel to,
you come with me.

But "now" is lost to us, a present
past sharing.

I turn to give you some small piece of news.
Infinitesimal, but you would have cared about it.
Even this atom is too huge
for nothing to grasp.

To the generations that will live a thousand years

Your coming is forecast daily in the papers.
Technology will tweak your genes
and you will dip test tubes
into that fountain of youth
we have always searched for, that spring
rumoured to rise in the land of flowers.

I am here to tell you, on this northern morning,
when leaves still run with raindrops
and a chipping sparrow
pours its choking stream of notes
over the lingering clink in eavestroughs,

that, however large a cup you dip
into the fountain, never will you stretch
this moment, now, a moment longer.
We can only drink it from our own
cupped hands, from which it drains
in quick, bright runnels—a few drops held
in the shallow nest of your palm.

Last scattering surface

The street-sweeper crawls
beside the curb. Its brushes blur
a whorl of winter dust,
to hide the far side of the road in brown cloud.

And I think of how
the "last scattering surface" hems in time—
a distant mist
at the farthest edge of telescopes

marking that sudden phase transition
when light separated
from matter's whirling brushes
to travel free.

Today, we have shifted phase out of winter
and light can play
through this cleared air
in a glad array of wavelengths—

caragana's plumy green, the rosy foam
of flowering almond,
the may-trees' curd and cream.
Spring's new matter.

While across the street, the sweeper's last
scattering surface
masks the receding past,
draws a line between

before (a blur) and after.

Let us compare cosmologies

1 Let us compare cosmologies

There is a beginning and a middle.
There is an arc of narrative.
There is a word, a large engraved initial.
There is imperative—
a cause, a god. Or not.
There is an end. A purpose.
Or maybe none. There is a plot
with reasons, reason. There is a circus,
a theatre stage of space and time.
There are equations at the bottom
or the top. There is a pantheon
of matter, motion, scattered photons.
And the questions every universe expects:
what came before? What happens next?

2 The Orphic follower

There is an arc of narrative
in which unaging Chronos stirred the broth
of pre-creation, poured it through a sieve
to separate bright Aether from the muddy fog
of Chaos. He shaped a silvery egg—which cracked,
and from its glowing albumen and globe
of yolk came tumbling gods, giants, witchcraft,
houses of the zodiac, scythes, snakes, odes,
devourings, bellies, caves, firmaments,
and divine souls salted with corporeal soot.
From the cupboard's mad phantasmagory
of mythological ingredients
          compulsively we cook
the whole mad, scrambled pan of story.

3 A pope

There is a word, a large, engraved initial.
This Big Bang Theory—its bigness
and its bang—could easily become official,
the pontiff muses. Its attractiveness,
theologically, lies in what it does
with light—oh, let it be!
From the physicists' first plasma fuzz
matter decouples—whump!—and light goes free.
This picture is remarkably consistent
with the dicta of that ancient document
whose words, thus far, have proven so resistant
to science. He'd like a rapprochement;
the gap is quite a nuisance. So this gift
of scientific *Fiat Lux* might make all fit.

4 The evangelist

There is imperative—there is an order
shouted in my ear so I can shout in yours.
There is a Will that justifies our ardour
and our presence on your doorstep with brochures.
The heavens are specified in finest detail
by the One whose purpose hides. We receive
his bulletins, delivered like an e-mail
sent en masse to those who will believe
rather than fret themselves to understand
the inner works. You need no more than this,
the little pamphlets, coarsely printed, from his Hand.
Here, take them! In obedience lies bliss.
The stars are fixed for you—all this is certain
and there's no need to look behind a curtain.

5 The philosophical skeptic

Or not. "The universe begins to look
more like a great thought than a great machine"
comments the astronomer. But who is thinking?

How do we know it's real—this meaning we impute
to redshifts, cold spots in the CMB, the stream
of data we're extracting

from our machines. There's nothing we can touch
or hold or really see. We only model
with conceptual cardboard—

and the arcane numbers that we crunch
may be more real than anything they label.
We've got it backward.

The universe is not creating us.
We're composing it, from thoughts and dust.

6 The nihilist

There is an end, but not a purpose
to everything—relentless outward drift
to vacuum. All effort worthless.
Utter destruction. Nothing will be left.
Even space sucked down its own gullet.
    All coherence lost to void
No remnant memory or will. It
looms—annihilation—unalloyed,
ahead of us, behind us, now. Our baseless
    base. We'll conclude
in a pointless, popping point. Meanwhile, aimless
    rage or lassitude—
who cares? Not the drifting galaxies.
Nor us, their meaningless debris.

7 The totalitarian

A purpose. Vaunting ideology
reduces every complex question
to the comfort of simplicity—
one appealing mantra, to be impressed on
the swirl and churn of nebulae.

Gravity is essential for cohesion.
It assimilates the isolated "I"
into compact collective. Accretion
  is how you get things done.
  One for all, all for one.

But the fist itself contracts with what it crushes.
Gravity compresses all its subject
matter, conflating justice with injustice,
thought with insurrection, to a nugget
  of paranoia, a black hole
  rotating round itself. Control.

8 The survivalist

There is a plot. We're stuck here in this island
universe. No one knows we're here, except
maybe some cosmic camera, part spy and
part voyeur judge. Jesus wept!
We need to get prepared. Lay in a store
of hydrocarbons—the tanks will have to be
enormous. Work out how to stave off meteors—
they're out to get us, and it's catastrophe
if they do. There are basic skills we'll need
but haven't practiced. Would you know how to make
a respirator for a planet? Or how to feed
yourself and fourteen billion? And we'd better pack
the bug-out bag—a little silver capsule
for a rocket's nose. We'll need a vehicle.

9 The optimist

There is a circus. Rings and hoops and bubbles
balanced by clever seals, applauding
their infinite flippers. Acrobats juggle
from the backs of prancing math, plumes nodding
as the ponies step through ring after ring
after ring in time. Streaming out behind
comes a froth of iridescent, multiplying
sapphire-tinted worlds, a multiverse of diamond
and black hole. Through its branching histories,
our doppelgängers live out all hypotheses.

And in the circus of the infinite
somewhere there's a system
where everything turns out all right.
It could be this one.

10 The magician

A theatre stage of space and time—
the curtains open. Bang. Their heavy velvet
swings through echoing ripples in the sudden shine
of light turned on full blast. Trussed and belted,
the magician is lowered into a tar-black vat
while the universe holds its breath—how
will he get out of this one? Then claps
and roars when he pops out to bow—
a huge inflationary surge of approbation.

The curtains will swing closed again,
the show conclude. But like the lady sawn
in half, it's just invisible contortion—
cunning illusion of severance and end,
a twist in a tiny space. The show goes on.

11 The baker

There are equations, recipes to follow
step by step for a universe to rise
from the gluten of gravity. You borrow
a cup of epsilon—that's to catalyze
protons into being. Then lambda—
just the smallest pinch, enough to lighten
the cosmic dough like baking soda.

If you mix ingredients just right, then
your universe will turn out perfectly.
It's all determined by the laws of physics,
inexorable cause, effect. Nothing is free
to vary. It's like a tray of biscuits
unfolding uniformly in an oven's bubble—
the results are inescapable.

12 The consumer

At the bottom or the top. We have
a universe from nothing, the ultimate
free lunch. We've done nothing to deserve
such luck—we sit back and accumulate
what we want from the vast conveyor belt
grinding past us, the continual assembly
of protons into elements that melt
from blown-up stars. We suck the energy
of the vacuum—that enormous teat
engorged and gushing over us,
a splashing cornucopia of milk and fruit
that we could never drain. Delirious
with plenty, we sit here in its middle—
kids at a seaside, an oceanic idyll.

13 The funeral director

Of matter, motion. So much emotion tied
to matter. Open your arms, release the corpse
to that faint whiff of formaldehyde
that's made among the stars. Time reabsorbs
all our bodies. Let them go.

        The loss is real,
        I know. You feel
all space has stretched awry, want to throw
yourself, your smallest molecules, after her
into the gap where she has vanished.
But know the universe is steady-state—matter
emerges out of starlight, undiminished.
We gain exactly the amount required to fill
the opened space, to make it bearable.

14 The Manichean

There is a pantheon. Beings born
out of the battle between two kingdoms
that have contended since before the dawn
of time. From the Kingdom of Cohesion
march its great commanders: Gravity, Strong
Force, Magnetism—and their myrmidons
of matter: gluons, electrons, protons,
who seek each other out across the universe
and bind themselves in one. The garrison
of the second realm, Oblivion,
is staffed more sparsely: Entropy, Vacuum,
and shadowy Lord Lambda—wan,
cloaked, master of rupture. These lines are drawn
from end to end of time, and then beyond.

15 The artist

What happens next? What happens always.
Creation. Not *ab nihilo*—we're not gods.
We compose the way a ball plays
off whatever nearby surface is the cause
of its trajectory. From random atoms
we make our lines. We listen for the echo,
the rumbling repercussion of what happened
before us. We let that set our key.

We magnify. The universe comes singing
out of its tiny throat of hidden space.
We take that filament of beginning
and loop it louder, spin a wilderness.
Then hand it out, our offering: *Here, view*
its subtle finish, try its bounce. It's made
for you.

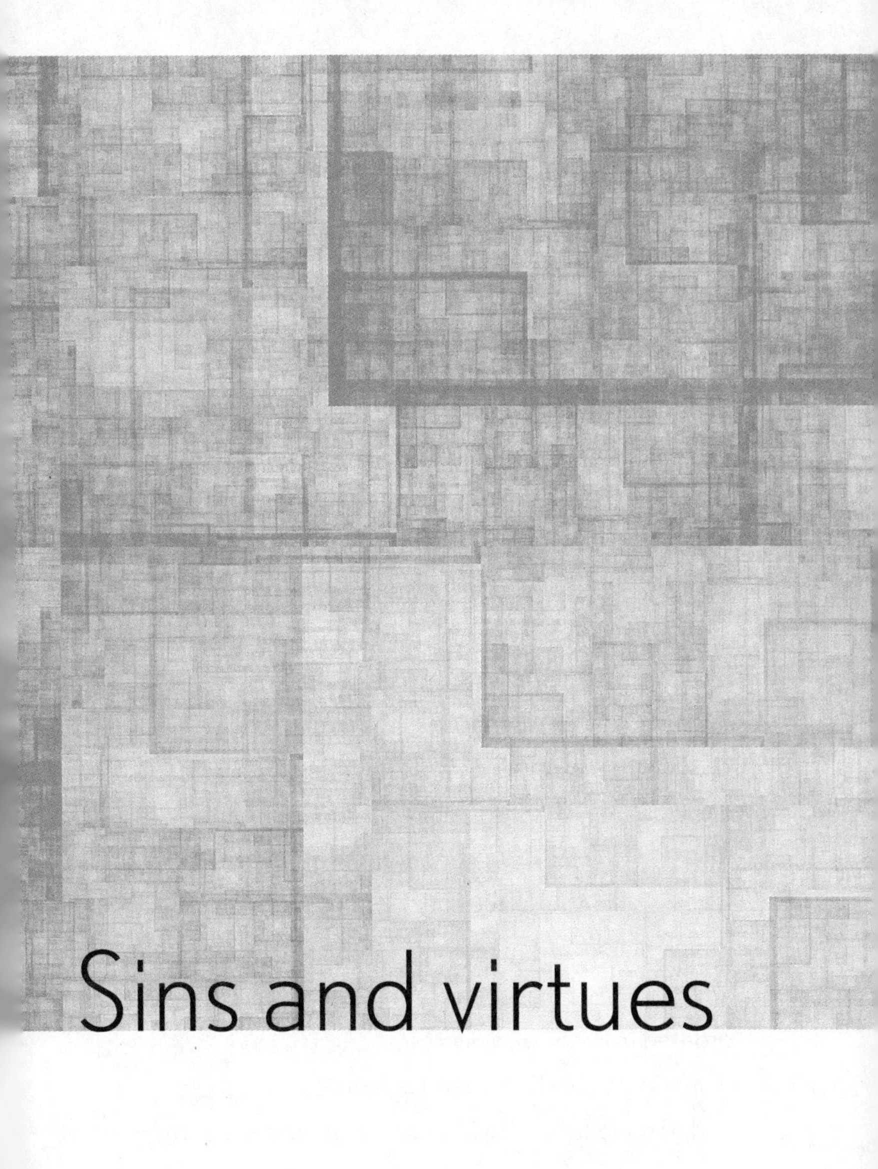

Sins and virtues

Avarice

Small and gnomish. The greeter at the door
of a great emporium whose shelves recede
deeper and deeper, stacked with more and more.
His vest flaps with credit cards, interleaved
like chain mail, and his hat is panoplied
with Christmas lights. He is benignly prompt
to murmur *Yes, it's here, it's what you need,*
we have it cheap. He leads you through the haunts
of acquisition, to the tawdry Throne of Want

which is piled with spoils from liquidations,
cheap labour, children's fingers, women's sweat—
the pit in which we dig our bargain basements.
Avarice's meagre silhouette
flits on ahead, luring us to get
any toy or trifle that contents us
ordered from the airy internet
as if its purchase had no consequences
and nothing further to be charged against expenses.

While the aisles of the emporium overheat
with demons, crazy-eyed and brazen-faced,
ripping boxes open, thrusting feet
in shoes they'll never wear, making a race
for T-shirts they will throw away. While *Waste*,

a florid figure gartered in the skins
of disappearing animals is braced
against a collapsing wall and grins
over this rubbled carnival of broken things.

Lust

Pornography's torpid ambassador
is a shadowy bulk, a faceless tower
that you might pass and hardly register,
were it not for that relentless square
inset in its abdomen. You have to stare
as if it was a TV in a bar; your will
to turn your head away is stripped of power.
Its fat printless fingers stroke the unreal—
the hot contortions we can touch but cannot feel.

Gluttony

It spreads—a huge and pulpy thumb, extruded
from a vast, obese fist. It presses down
upon the scales, its straining joints occluded
by swathes of fat, forcing the toll of pounds
to climb ever higher. The fist surrounds
whole populations in its podgy wall—
a catastrophe of bloat that mounts,
battening on appetites evolved
to savour rarities now rendered cheap as salt.

The thumb weighs down the price of fat and sweet.
It flattens the cost of corn, to subsidize
the fist's unbridled butchery of meat.
Gluttony's bland ministers supervise
a syrup milked from grain that multiplies
sugar into sickness and addiction.
They trim fat only from their costs, supersize
the profit margins in their sizzling kitchens,
drive burger-flipping imps with sales predictions.

The thumb's weight presses heaviest on those
who should be smallest—children, lured
through the archways of the adipose
by piping advertisements and immured
in fastnesses of fat—castles toured

by laughing parties led by clowns whose grins
are painted round enormous mouths. The poor,
who have not time or coins for better things
to nourish them, are given toys and welcomed in.

Envy

Envy is that tiniest of things:
the worm that lives below the eyelid
of the hippopotamus, festering
and feeding on its tears—a meal provided
by the bevies of svelte beauty, sky-clad,
in magazines. Such pictures liquidize
our solid self-esteem, which has collided
with the knowledge of our lumpish, lumbering size.
The worm takes nourishment from our infected eyes.

Pride

A tank, titanic, its armoured shell
thicker than ignorance, its metal cast
in the megalomic foundries of hell
from pits of slag and ore. It trundles past
bristling with gun barrels, its treads crevassed
with broken limbs. It grinds with anthracite
conviction; its blinding crests of brass
and gunmetal dazzle those who might
contest its obdurate presumption that it's right.

Right. Right. Right. Somewhere inside
the confines of this iron firmament
a centralized authority resides,
controlling levers, aiming armaments.
Legions of demons, obedient
to arbitrary orders of attack,
bombard. Bombast. It brooks no argument,
no alternative analyses of stats
or proofs. *Do not confuse me with [those other] facts.*

Right. Right. Right. The right to level war,
to roll down opposition, stifle views.
No need to question what the battle's for
rather than against, or whose right is whose.

Gun turrets swivel, but *Pride*'s treads refuse
to vary its direction. Rhinoceros
with a single horn, determined not to lose
regardless of whatever may be lost.
Right, whatever may be left. Right at any cost.

Anger

The face of *Anger* flakes, repeats, repeats,
fractal and repetitive as fish scales.
Its eyes stare out at us from screens and sheets
of newsprint. We study their details
in the aftermath of tragedy, but fail
to find that something different in the pupils—
that brutal mark of Cain, however pale,
that must be coded in the grainy pixels—
something we missed, an omen warning us of pistols.

Behind those eyes, the entrance to a chamber
filled with rifles that repeat, repeat
the burden of their arsenal of danger
into our horrified hearts. A vault heaped
with oiled assault machines to aid complete
annihilation. It should be shut and locked.
But *Ideology*'s recoil defeats
the safety catch and *Profit*'s toe is caught
in the door. And so the demons spiral, cannot stop.

Sloth

Sloth's mesh hangs wide and high—immense net
fencing its detainees around. Their fingers
trace its strands and forked diversions, get
distracted by its branches and the stringers
that lead to time-dissolving ends. They tinker
with clicked connections that disperse
all purpose. Hyperglycemic linkers
lured by easy sugars, they are immersed
in the coils and convolutions of a multiverse

that's one dimension lower than the real—
a hypersurface only, an extraction
from that higher, deeper space concealed
behind this web of interlaced attractions.
The mesh is manufactured by the passions
of Sloth's engaging troops, because incessant
industry is needed to advance inaction;
they volunteer their huge investment
of time to losing time inside this idling present.

Sloth's imprisoned minions crawl the walls
of their labyrinthine honeycomb—
a waggle-dance exploring the ephemeral
with clicks and programmed links to home
pages and buzzing social media. Alone

at the screen's mazy face, they feel busy
as worker bees. While *Sloth*, a queen enthroned
on honeysuckle pillows, fans her gauzy,
droning wings to air the hive and keep sweets easy.

Mercy

Mercy is a blue clay ball, revolving gravely
beneath a slip of white gauze. She is matter
distilled from the infinite dark
that envelopes her. She is a lovely,
scrupulous clerk

who keeps exact account of what is owed to her—
a tally of the ores and atmospheres,
the old petroleum
and soiled soils that are clawed from her.
In her scriptorium

a recording angel with a face of silver
—a battered attendant moon—
rolls up a lengthening scroll,
an annotated list of what's been pilfered
and what's been sold.

Mercy's debtors sense their vulnerability—
that there will be reckonings.
They do not quite forget
that *Mercy*'s bonds are those of loyalty.
No clause or caveat

wipes out the fact that *Mercy* holds the upper hand.
The circling angel twists the scroll again.

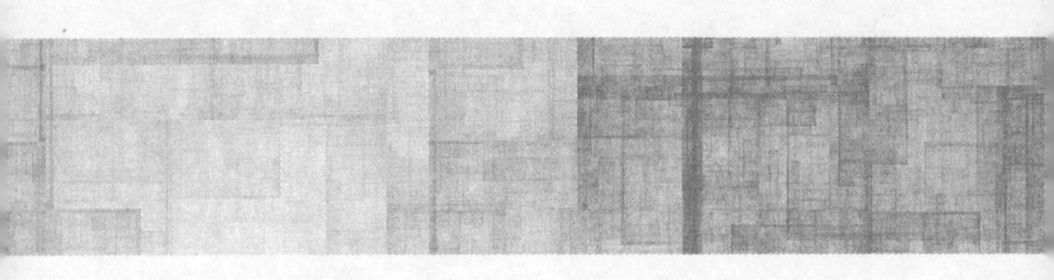

The debtors know, in fairness
we deserve the dark. Will she demand
her bonds? Or spare us?

Hope

You come upon it suddenly
in some abandoned square
reached by random side-streets
that you forgot went there.

A statue in stolid granite.
It is carved life-sized—
neither more nor less—a girl
with ordinary eyes.

At first you do not notice
the draping of a cloak
stitched with a thousand feathers,
knotted at her throat.

A sheer, grey, unseen blur—
unless, perhaps, the light
falling athwart office towers
catches its faint weight

and stirs it to iridescence—
radiance afloat.
Hope is that thing with feathers
catching at the throat.

Shifting wavelengths

Tortoise and fern

"It's turtles, turtles all the way down."
—creation myth humour

No, not just turtles. It's the curling alternation
    of tortoise then fern.

Galaxy's carapace turns slowly.
Starry claws pull it through its sparkling circle.

    Orbits of planets effloresce, draw fronds—
    lines traced by wands with silver tips.

Earth's enamelled shell, plating of air.
Polished layer of porcelain.

    Forest's scrim-edged coastline, like patches of lichen.
    Elaborate garden on stone.

Trees' canopy. Domed shelter.
Hearts at home in its shadowy interior.

    Veins unfurl, fine into finer. Each junction finds
    a suitable angle for branching.

The terrapin self-containment of cell—
ancient invention of boundary.

Crinkled spring spiral of chromosome
uncoils its maidenhair cascade.

Molecule balls.
Atoms corralled, held in thrall to edge.

Spray of gamma ray turns a cloud chamber
to a fern-filled terrarium.

Quarks. Trapped triolets, smoothly edged
as three-leafed clover. Never four.

Quantum foam unfolds its fractal sparkle
where galaxies can swim.

Fingers of God

It's an optical illusion—long strings
of galaxies that point directly at us
in telescopic images, fingering
our privileged and central status
in God's infrared field of vision. But
it's just an accidental artefact
of light's red-shifted waves—like the effect
of driving into snow when every flake
seems aimed at you. Make no mistake,

don't think they care about us. In fact,
they've stuck a key in their ignition, stepped
on the gas, stick-shifted away from us.

They're neither indicator light nor portent.
Get over it. We're not all that important.

The barber's paradox

The Barber's Paradox is very bad for business.
"I'm only going to shave you if you never shave yourself,"
says the sign on the shop door, below the striped pole
where red and white revolving lines chase themselves silly
into spirals. "But it's Saturday," protests the customer
whose beard bristles insistently through his skin.
"I'd like a little luxury today—don't usually have time
for nice hot towels and that lather on your swirling brush
scented of coconut and lime."

But the barber is implacable—he will not shave
those who shave themselves. He is a man of principle.
His own beard foams across his chest, billows
over the shiny crimson leather of the chairs, fills up
the shop, because of course, he cannot shave
any man who ever shaves himself. The customer
grouses. "Well anyway, you'd never find the razor
under all that hair." The bell tinkles. The spiral pole revolves.
The barber, somewhat sadly, thinks it's hard to pay the rent
when your theories are set.

Zeno's paradox

We've solved the paradox.
Motion is possible. The arrow's flight ends
even if its fractions interlock
to infinity—half a distance, yet again
half, and half .... We know this series sums
to a finite thunk and shudder.
And we know
the thrumming calculus of life comes
to completion. I am half-way through
my count of years—half-way to knowing
all I will know. Yet something stalls
in the air, an infinitely subtle slowing.
Of whatever I have learned when the arrow falls
silent, one last sliver will be lost.
A final distance will remain uncrossed.

Twin paradox

I am your twin. You, the starry-eyed young woman
in the heart department, newly engaged, exploring
the curvature of future at light speed,
wearing your fine skin stretched
on its delicate frame.

It is as if you had been accelerating in a rocket ship,
travelling to arc around a distant star
and returned to find me,
your twin, with time
puddled at my toes.

While your clock was stretching out its tick-tock metronome
I was making all those astronaut decisions
that swing open now before you
like the hatch of a landing capsule
on an unlived world.

Your young face reflects mine, but our mirror symmetry
is wrinkled by time's stretch marks. You greet me
on this green-again planet
without recognition.
That will come.

Sand reckonings: Eubulides' paradox

How many grains of sand are needed for a heap?
One is not enough, or three.
Where does *small* become *large*? A beach
has many grains—we will agree
that number's large, however small the stretch
for spades and pails and castles.

Sand hundreds. That's the name the ancient Greeks
coined for the uncountables.
But any number's small when we compare
its span (however vast) with all
infinity beyond it, numberless and fair
with castles we can never reach.
Sand hundreds trickle quickly through
my fingers—never days enough for beach
and you.

Honeycomb conjectures

Hypothesis 1
Wax's fragile scaffold
can bear the weight of pounds and pounds
of sweetness.

Hypothesis 2
Repetition is acceptable.
Honeycomb. Hexagon. We could tile the universe
unending.

Hypothesis 3
Wax's mathematics. Six
straight sides contain least wax,
most honey.

Hypothesis 4
Wax accumulates.
Tiny beads exuded, glandular secretions,
excretions, accretion.

Hypothesis 5
We can be fed into existence—
sliver of silver egg inserted into honey—
then worked hard, hard, hard.

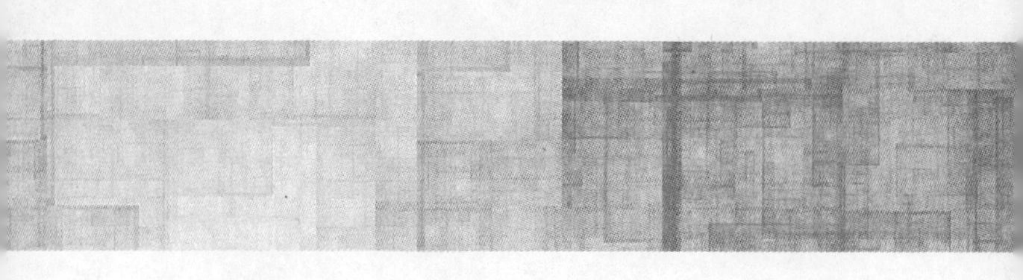

Hypothesis 6

It takes a million flowers
to fill our honey stomachs. A world sucked up,
squeezed out.

Bee violet

What is this colour we cannot see
that marks a cross
at the flower's nectared centre?
So conspicuous
to your hive of velvet saints
who then exhaust
their busy wings upon this
luminous boss
of bee violet nailed to the blossom—
a vivid gloss
upon the petal's scripture
invisible to us.

Optical molasses

Winter's light approaches us,
blue-shifted, gelatinous—
a Doppler contraption that will herd us
into slower and slower motion,
a condensate of cold weather.
Under its needled ether
we contract, huddle down together—
atoms nearly frozen

into a syrup where the individual
is rendered indistinguishable.
But this down-parka-puffy-oneness
is fragile, fractures easily.
Molecules collapse,
our pulled taffy
snaps.

Life adapts to inhospitable environments

Snowflakes sidle in the air,
sparse paramecia in a test-tube solution.

Ice has packed down on the roadway, hard
as the glass below a microscope's ground lens.

Apartment balconies are stacked to the grey ceiling,
like drawer handles in a naturalist's collection.

Is the eye clinical or kindly? Regardless,
in this harsh niche, the cell life multiplies.

A bent old woman's cane wavers before her,
agitated, energetic flagellum.

The man in the red toque rolls a cigarette
in one clawed hand. Ash fragments flake.

Specimens impenetrable as curios,
transparent as a slice of stained tissue.

How to tell a Martian my heart is on the left

It has to do with magnetism—
how electric current wraps a magnet's heart.
It's that direction.

It has to do with parity—
how it's not conserved, how fate is not
shared equally in weak decay.
It's that direction.

Asymmetries appear to be
so deeply universal. They do
get things going,
pump currents of economy
into a static world,
set its direction.

Yet there are consequences
when things are not shared equally
and, dear Martian, I cannot help
how I am made. All these electrons
hooked together in my core,
most of them spinning to the left.
Ten to however many powers of them
hauling my politic heart
in that direction.

Underworlds

Persephone and I are underground

waiting through winter, gazing out at icicles
a ragged frieze
of glass roots growing downward
from the eaves.
Some tapering and thin. Others
gnarled as carrots
that have encountered stony soil.

We bear it,
attired for this realm in padded parkas
and stiff-toed boots,
our fingers smoothed into mittens, our heads
rounded by hoods.
And blunted into earthworm shape
we thread our way
through cold's stiff overburden—corridors
of stubborn clay
that we negotiate, rather than confront.
Above our heads
a mesh of barren branches tangles
fibrous webs,
an airy turf that reaches down for nutrients.

We are surprised,
somewhat, to find ourselves still here,
though we surmise
that it is somehow due to love.

Not for the men, those fierce accidents
who brought us
down from the fields of girlhood,
free and thoughtless,
and hold us here. No, love has become
something we hold
in mittened palms: a scattering of arils
red as blood.
They are faceted, irregular, translucent
as garnets mined
from crimson veins in granite. But cased
in softer rind,
and freed now from astringent pith.

We seem to be
responsible for them. In this alien
geography
these seeds are to unfold and germinate.

Tiny radicles
should probe the fine-tilthed soil in pots along
our window sill,
their cotyledons lift. But here's no climate
for pomegranate.
How do we render such a tender thing robust?

Still, planted
they must be, and sheltered, raised to fruit.
Persephone
and I bend our heads together, puzzled.
What husbandry
can make this happen? Why this task?
Why here? And who has asked?

The outer dark

Downtown mall—the dark end
of the pedway, where sunlight
doesn't reach. Footsteps part and bend
    around a clear-walled case, the height
    and width of a tall man. Within,
    flickering from a bed of glassy quartzite
pebbles that shine like glycerine,
gas flames writhe and float—
a cold fire burning nothing, a skin
    of warmthless heat, remote
    and unconsuming conflagration
    within its crystal envelope.

Mall management has taken
the suite of leather-covered chairs away
that used to seat the weary at this station
    along the trodden passageway.
    A wide bench of blonde wood
    circles the hard hearth. Seated there,
his back to the great glass cube
is a young man's slumped figure.
His face is shrouded by the hood
    of his ragged jacket—sizes bigger
    than his skinny torso. Cuffs pulled long
    over knuckles. Cowled, hunched, meagre.

How do I know that he is young?
He could be ageless, a grey friar
of infernal realms. He might belong
to the legions of fire,
the imps of darkness, stamped
with the cloven hoof of gang attire.

The mall guards would think so—vigilant
and righteous, they watch out
for the insignia of such miscreants
and interrogate them, like devout
and diligent inquisitors
sternly seeking demons to cast out.
And yet they leave this visitor
unapproached for now. Perhaps
they are simply busier
somewhere else, sealing up the gaps
that let in sulphurous air,
on orders from the bureaucrats
above. Or perhaps it's fear
of what we'd see, a dread
should that mute figure of Despair
ever raise its head.

I am seized by a strange, crazed fancy—
If he looks up, I am afraid
that blackness will look out, a vacancy
more utter than the icy depths of space,
that I will stare into that hood and see
a faceless face
that convicts my world as nihilist
and empty—a stupid, useless place
to be dismissed
to the furthest fringes of the dark,
not worth the slightest twist
of a telescope. If Despair looked back,
held up a mirror from his anteroom of hell,
the reflection might be blank

and this whole world might disappear as well.

Cocytus

Inner city, winter dawn—
blocks of battered buildings, ringed by towers
and shadows of the towers still to come.
The giants of development lour
over this landscape's rigid lake
of ice. In this pre-daylight hour
at a season when it seems no light will break
ever, the streets are in the lock-down
of cold. Nothing should be awake
or moving. Yet figures mill around
the doorway of the drop-in centre, where
an ambulance's flashing lights resound
from ice-glazed surfaces. Officers
of medicine in sombre uniforms
are coming down the stairs,
the faces of professional concern.
Followed by a social worker, forehead
grim with worry. She turns
to a colleague with the horrid
information: "It looks as though
he might lose all of them." The florid
excess shocks: *all* ten fingers lost to snow
and frostbite. Two reaching hands
gnawed to clubs. Nausea and vertigo

churn stomachs. How can a man
ever seize his life again like this? How shitty
does it get? How do we understand
this—in the middle of a city
lined with warm houses? Where brute
want is soldered to prosperity?

They've seen it all, here: A gangrened foot,
putrid, that comes rotting away
when a man pulls off his boot.
Another man who waits beside the entryway,
palm held out, and swallows
anything that's dropped in it. Hey,
whatever pill or capsule that his fellows
toss to him might halt the pain
and make the constant roar of sorrow
shut up. Sure, he's eating his brain
alive. But where else will he find relief,
however fleeting? We think that shame
should shape you up—the old brimstone belief
in the eyes of judgement. But when days
are one unending pit of shame, the grief
will drive you to the worst, crazed,
trap-door of escape. Dante had it right—
in hell, you do again the thing that shames

you most. Now the ambulance's lights
go silent as its wheels depart
across the frozen surface of Cocytus.
Another morning starts.
The drop-in clients gather round the entry
with their battered shopping carts
while drivers on their way to heated plenty
in parkades below surrounding towers
frown at their idleness. *Tsk, tsk. Empty*
beer cans on the sidewalk. It puts our
property at risk—I say
that centre's just a magnet. Powers
and commercial principalities convey
dissatisfaction into city halls.

True, these people are not innocent. They betray.
But who more than themselves?
Who's most hurt by the crap?
The petty mischief, the small-time deals
done to survive, to cope.
They are driven to devour each other's poison,
cannibalize their children. They are as trapped
as Ugolino in his starving prison.
Can those with nothing care about your property?
Hear the wind blowing. Will you listen?

The terrible archangel of poverty
is frozen into this lake of ice
up to his waist. His wings whirr endlessly,
        whip up the windy blasts of Boreas
        that, channelled by the towers, bring
        creatures to their knees. At this bitter impasse
the judged and shamed can only cling
to his hairy haunches, unwanted pests.
For them, no exit from this suffering.
        The city scratches at their bites, arrests
        their bodies, not their pain.
        No exiting these gelid depths
for anyone.

Niflheim

This is a statue to the memory of
the unremembered,
the homeless ones who die
(thirty, forty
of them) every year in this one city.
Thirty or forty
names carved only on the air by frost's
fleeting glitter.

A slumped bronze figure sits
by a closed door.
Through the door frame's metal rim, we see
no light, no fire.

The door that opens for those who die
without a home
is the door to Niflheim, the realm
for those cut down
by age or sickness, a land of mists
and misery,
locked behind Hel's heavy gates.
("Old" means fifty
in the inner city. Disease as prevalent
as winter weather.)

Around the arch of this memorial
square tiles of clay
record the thoughts of those who know
what "homeless" means.
Spare some change, in raised block letters
carved by one
who knows how little change occurs.
Another winks
No diving beside a sketched dumpster—
thumbing the nose
at authority's directives. A heart
has *Hope* inscribed
on its rounded surface, sheltered
in curving palms.

Even the homeless memorial
found it hard
to find a home. No, not on the civic
squares and plazas,
declared authority. No space in front
of city hall.
We do not like to think of Niflheim.

So here it sits
in the realm of old railway yards
and redevelopment.

No cleared pathway to approach it by.
Surrounding snow
collapsed to hardness. On this
January day
it's like walking on a choppy sea
modelled in hard
glazed pottery by hurried hands.

Heroes earn
the warrior's end, Valhalla. They cross
the rainbow bridge
to feast and plenty. But those who die
of sadness
reach Nifelheim across a shore of corpses,
and their battles
against the giants go un-named,
unrecorded.

The man with no hands

The man with no hands stumbled in
from the social service labyrinth
to your office. Another of the men
who negotiate the nether regions.

He was mumbling, incoherent—
though not just drunk or stoned. *I've never seen*
anyone so tired, you say, amazed
that exhaustion could ever be so palpable,
the special hell for those who have no place at all
to rest, no cave or refuge. He fell asleep
in the chair beside your desk, while trying
to tell the tale of where he'd been
and where he had to go.

You let him stay there, sleep the only gift
that you could give him then, or he could take
with those maimed hands. His prosthetic claws
lay beside him on the floor, clumsy hooks
that chafed the stumps of flesh they clamped to.

Those who travel to the underworlds
lose pieces of themselves. Orpheus, Osiris,
Persephone—none return whole.

He slept in that hard chair for hours.
You left him briefly for a conference,
returned to find him gone, and with him,
his dead hands.

Each of us the centre of a circle

You ready yourself beside the door
—keys in one pocket, cell phone in a second—
to enter this fresh morning. You have dressed
with unaccustomed ceremony
in garments from the corporate armoury,
the tie-and-jacket that you wear so seldom
now. Such protection scarcely needed
for the daily dangers and alarums
of your inner-city office. But today
you are encountering the plated uniforms
of police and politicians who desire
to "clean up" the troubled people
who intrude upon their plans for order.

Yesterday, you called me out to see the arc
of a rainbow—a radiant circularity
against the dark pewter polish
of the storm just passed. The clouds' livid canvas
swashed with nested bands of colour.

Although our shoulders touch, we each inhabit
a different rainbow, whose wavelengths travel
unique and separate radii to our eyes.
Yet something indivisible still overrides
the tiny separateness between
our overlapping arcs. We register

the same intensity of wavelength. My rainbow
spans the same extent of sky as yours.

This is our shining armour—
a shared spectrum, the colours' constant order,
your polished shoes.

Postscript

God submits a grant application to the Canada Council

Purpose of the project: *To create a world.*

Detailed description: *In this multi-disciplinary, cross-genre work, I intend to create a self-sustaining performative experience based on the geometry of the sphere. It will draw on my unique background in arts and theoretical mathematics, incorporating prose and poem with theatrical techniques.*

God considers what to add.
She knows she wants her world to be
round, and breathing. Movement, too—
the whole thing should shimmer
and shift. She has a picture in her mind
of great indeterminate beasts
shuffling through some waving stuff.
Sounds, as well—booms
and crashes, sussurations.
Something that goes *hiss-sss-ssss.*
Can she manage smell? How would that
work? Some medium to float
triggers in, perhaps, and then
a tissue of receptors
to twinkle at cinnamon. Yeah,
she could come up with something.

The project will culminate in a performance lasting seven days. Given the potential audience interest, I feel confident the project can travel and be repeated in other venues.

Oh, geez, she'll have to find
an audience somewhere. Need
to think of that.

Budget: *Subsistence, twelve months at $1,900. Materials, nothing. (She'll work with what she's got lying round out back.) Venue rental: $2,000. Miscellaneous: $200 Total $25,000. (Might as well ask for the max.)*

God sends the application in
and waits. The jury process takes
eternity. She keeps noodling
on the concept, realizes
it's going to take a whole lot longer
than she thought, to conjure
"theoretical" into "applied."
How does she get everything to stick
on some slippery cerulean ball?
She'll need to create some force
that grabs a hold. And then that bright idea
about receptors—they're trickier by far
than she ever thought at first. What the hell
is she going to mount them in?

She half-decides to let the idea go.
She could do something simpler with a foam,
maybe. You'd get some nice effects
just swirling that around. Leave words
out of it, forget the "prose and poem."

She's forgotten where she put her notes
when the envelope appears.
The grant has been approved. "Oh, shit,"
she thinks. "Now they're expecting it."
She sighs, dispirited, then peers
at the amount she's been awarded.
Sure they gave her what she asked for.
But how is she ever going to live on this
for fourteen billion years?

Notes

The set of all gods

The god of gravity

Gravity is the weakest of the four forces that hold matter together. However, it operates over a very long range. It is always attractive (unlike, for instance, the electromagnetic force, which has positive and negative charges), and acts between any two pieces of matter in the universe however far apart they are.

The god of kites and darts

Roger Penrose calls the two shapes of his famous tiling "kites and darts." Fitted together, these two angular shapes will cover a plane infinitely, but the design will never repeat exactly.

The baker god

Cosmologists speculate that the multiverse might throw up many different universes with different defining characteristics—we just happen to live in one that has the right conditions for our world to form. (See *A Universe from Nothing* by Lawrence Krause.)

The god of hearts

Jupiter Tonans was one of the classical epithets for Jupiter in his role as god of thunder.

The jeweller god

In white dwarf stars of the right size, the carbon atoms will slowly crystallize into a diamond lattice as the star cools.

The god of dark

"Dark energy" is a hypothetical form of repulsive energy invoked to explain recent observations that the universe is expanding. It permeates all of space. (See *The Lightness of Being: Mass, Ether and the Unification of Forces* by Frank Wilczek for a very clear and interesting discussion of the fundamental components of "empty" space.)

The muse of universes
This poem takes off from "brane theory" and the idea that the Big Bang may be born from the repeated collisions of two branes that collide, slowly separate and then re-collide. (See *The Universe Within: From quantum to cosmos* by Neil Turok.)

Ordinary matter

The helium thoughts
Our sun burns by nuclear fusion, turning hydrogen into helium, which are the two simplest elements.

Love in three dimensions
The article "Leaves, Flowers and Garbage Bags: Making Waves" (by Eran Sharon, Michael Marder and Harry L. Swinney) describes the fractal symmetry-breaking that creates ruffled petals and lettuce leaves. The ruffles are the simplest way the necessary buckling can happen in ordinary three-dimensional Euclidean space. (*American Scientist*, May–June 2004.)

Local bubble
The Local Bubble is a region of very sparse interstellar gas through which the sun and solar system are currently travelling. It seems that the "cavity" was formed by supernovae that exploded within the last 10 to 20 million years.

Standard candles

Clouds of glory
Henrietta Swan Leavitt was an American astronomer who worked at the Harvard College observatory under the direction of famed astronomer Edward Pickering. Her breakthrough calculations, based on her analysis of stars in the photographs of the Magellanic Clouds, revolutionized astronomy. The Milky Way could no longer be viewed as all of the

universe; instead, it was one of many "island universes." However, she was given little credit for her discovery at first and very little is known of her life. Her story is told in *Miss Leavitt's Stars*, by George Johnson.

The blink comparator is an instrument that allows photographs of the same star groups taken at different times to be superimposed on each other, so that differences in star magnitude can be spotted more easily.

Pythagorean theorem

The Pythagorean Theorem relates the length of two sides of a right-angled triangle to the length of the hypotenuse—the side opposite the right angle—by the square of those lengths. It is one of the oldest geometric observations in mathematics and foundational to many mathematical concepts. Its central equation ($a^2 + b^2 = c^2$) can be extended to calculate distances calculated in three dimensions and higher.

Triangulation

The Great Arc by John Keay tells the tale of the Great Trigonometrical Survey, the monumental task of mapping the Indian subcontinent. The survey was begun by William Lambton in 1802 and completed by George Everest in 1837. Lambton was driven less by the ordinary colonial needs of mapping than by the ambition to determine the exact curvature of earth along an arc of longitude, and thereby the planet's shape.

The end of greatness

The observable universe is made up of a hierarchy of structures (galaxies, filaments, walls, etc.). However, at a scale of roughly 300 million light years, these structures no longer seem to "clump" into still-larger ones. From this scale on, the universe seems to be homogeneous.

In the Castle of Stars

Stjerneborg was the name of one of two observatories built by Tycho Brahe on the island of Hven, which had been granted to him by King Frederick II of Denmark. As a younger astronomer, he had made careful observations of a very bright "new star" (stella nova) that appeared in 1572. In his book outlining the observations, he was strongly critical of those who tried to dismiss the radical implications of the new star for traditional Aristotelian cosmology. ("Oh thick wits!")

"His celebrated metal nose"—Brahe had a good part of his nose sliced off in a duel when he was a young man, and for the rest of his life wore a metal replacement.

Supernova Type 1 A

Professor Gerson Goldhaber joined on the Supernova Cosmology Project at Lawrence Berkeley National Laboratory, after a long and distinguished career in particle physics in which he helped identify the signatures of several new subatomic particles. The team was formed to use observations of supernova to assess how fast the universe was expanding.

Supernova Type 1 A makes a useful "standard candle" since its physical processes require stars of a consistent size, which makes them remarkably uniform. Therefore, any differences in the measurements of brightness and wavelength in the radiation from such supernovae reflect how distant they are and how fast they are moving away from Earth.

A histogram is a graph that represents how statistical data is distributed—for instance, the higher a column, the more examples of the data fall into that category.

$d = (X-x)^2 + (Y-y)^2 + (Z-z)^2 - c(T-t)^2$

This is the formula for calculating the distance separating two events in space-time, according to Einstein's theory of relativity. It extends the basic Pythagorean theorem ($a^2 + b^2 = c^2$) from the two dimensions of a triangle drawn on a flat surface up to four dimensions. In Ian Stewart's *In Pursuit of the Unknown: Equations that Changed the World*, his chapter "Relativity" helps explain the relationship.

Muscle of difficulty

Muscle of difficulty

Charles Darwin called the corrugator "the muscle of difficulty" in his book *The Expression of Emotion in Man and Animals.*

Yet another crack in the foundation

The phrase "land of moles and pismires" comes from Thomas Browne's essay, *Hydriotaphia, Urn Burial, or a Discourse of the Sepulchral Urns lately found in Norfolk*, published in 1658.

The movers' dilemma

This problem is described by Eli Maor in *The Pythagorean Theorem: a 4,000-year history*. "A moving company needs to carry a long object, say a sofa of length a through an L-shaped hallway. Will the sofa get through?" The answer is related to the geometry of triangles and dimensions such as the width of the hall.

Rectangularization of the morbidity curve

In health statistics, the morbidity curve plots the percentage of people in each age group who do/don't experience chronic health conditions. If the average age of becoming "infirm" can be raised, this would theoretically square off the curve and shorten our years of "morbid life."

Last scattering surface

The Cosmic Microwave Background radiation is the faint leftover radiation from the point in the history of the universe when light and matter separated, allowing protons and electrons to combine into matter and photons to travel free. The CMB radiation appears to come at us from a distant "shell" around the observer, representing the distance each photon has travelled since that fundamental phase shift. The surface of this shell is known as the last scattering surface.

Let us compare cosmologies

The Orphic follower

Orphism originated in ancient Greece. The texts associated with it are ascribed to the mythical poet, Orpheus.

A pope

In the mid-twentieth century, Pope Pius XII was intrigued by the idea of the expansion of the universe from a "primeval atom" (later named the Big Bang theory). He saw the theory as a potential marriage of science and religion and justification of the Creation narrative in Genesis. However, he was warned off adopting it by the astronomer, physicist and priest, Georges Lemaitre, who had first proposed the theory but felt that religion and science should stay separate.

The philosophical skeptic

The English astronomer Sir James Jeans wrote, "The universe begins to look more like a great thought than a great machine."

The optimist

Some cosmologists posit a "multiverse" from which different universes (like the one we live in) emerge. In *The Hidden Reality: Parallel Universes and the Deep Laws of the Cosmos*, Brian Greene explores the various lines of mathematical thought that lead to different kinds of multiverse.

The magician

A number of cosmologists like Roger Penrose think that, rather than emerging from nothingness at the Big Bang, the universe is cyclic, with a new universe emerging from the end conditions of the old one. (See Penrose, *Cycles of Time: An Extraordinary New View of the Universe*.)

The baker

Epsilon and lambda are two of the basic parameters defining our universe. Epsilon (ε) is the finely tuned percentage of the mass in an helium nucleus that can turn into heat when it fuses. This enables atoms to fuse and heavier elements to form. Lambda (Λ) is the energy density of the vacuum; its value keeps space itself expanding. (See Martin Rees, *Just Six Numbers: The Deep Forces that Shape the Universe*.)

The consumer

Physicist Alan Guth, who founded the theory of cosmic inflation, said that "The universe is the ultimate free lunch." (See Stephen Hawking, *A Brief History of Time*.)

The funeral director

Formaldehyde was the first complex organic molecule discovered by astronomers in the interstellar medium. Its distribution is widespread within our galaxy and beyond. The steady-state model of the universe has been proposed as an alternative to the Big Bang; it is based on the hypothesis that pockets of new matter are created over time to compensate for the universe's expansion.

The Manichean

The Manichean religion was founded by the Persian prophet Mani and based on a fundamental dichotomy between dark and light. The elaborate cosmology described an ongoing struggle between the opposing forces (rather than a single all-powerful deity.)

Sins and virtues

The stanza form used in this section for the allegories of the seven deadly sins was invented by Edmund Spenser and used to create his great work, *The Faerie Queen*. In Book 1, Spenser introduces the Seven Deadly Sins as allegorical figures.

Envy

Oculotrema hippopotami is a minute flatworm that lives below the eyelid of the hippopotamus and feeds on its tears.

Mercy

The word "mercy" derives from the same root as "mercantile" and "market." It was invented by sixteenth-century translator Miles Coverdale to convey the Hebrew word *chesed*, which has associations with concepts like "tie," "agreement," "loyalty."

Shifting wavelengths

Tortoise and fern

Stephen Hawking, in *A Brief History of Time*, tells the story of the little old lady who tells a distinguished scientist that the world is really a flat plate on the back of a giant tortoise. "So what does the tortoise stand on?" asks the scientist. She responds, "it's tortoises all the way down." The phrase "turtles all the way down" has been adopted as a humorous way of expressing the problem of infinite regress.

Fingers of God

Red-shifted light from distant galaxies makes them seem like elongated fingers pointing back towards the observer.

The barber's paradox

This paradox is a common-language version of one that exposed a huge problem in set theory and changed the course of twentieth century mathematics. It has to do with whether a set can be a member of itself.

Zeno's paradox

The philosopher Zeno lived in the 5th century BCE.

Twin paradox

Einstein developed a thought experiment about two separated twins as a logical extension of his work on special relativity. Clocks that are moving with respect to one another will find that they have calculated time differently when they are brought back together, due to the relativistic effects of accelerating through space-time.

Sand reckonings

Eubulides of Miletus (4th century BCE) posed his "sorites" paradox, which can be articulated as follows. Premise 1: a grain of sand is not a heap. Premise 2: adding a single grain to something that is not a heap does not turn it into a heap. So there's no point at which a heap forms. *The Sand Reckoner* is the name of Archimedes' book on how to count large numbers.

Honeycomb conjectures

The honeycomb conjecture (first posed more than 2,000 years ago) was proven in 1999 by Thomas Hales. The conjecture states that the hexagonal arrangement uses least amount of wax to enclose the most space.

Bee violet

Certain flowers have "nectar guide" patterns that reflect ultraviolet wavelengths of light that bees can detect.

Optical molasses

This is a technique of cooling atoms down close to absolute zero until they become a condensate in which all the atoms are exactly the same. However this state is very fragile and hard for lab technicians to maintain.

How to tell a Martian

Physicist Richard Feynman used the challenge of trying to explain the directions "left" and "right" to a Martian as a way of exploring the laws of symmetry in physics. The weak force, which holds electrons and protons together inside atoms, is biased towards "our" left when atoms decay.

Underworlds

Cocytus

In Dante's Inferno, Cocytus is the lowest point of the circles of Hell, a lake of ice. There are several pronunciations, but for the purposes of this poem's sounds, think of it as Ko-Kite-us.

Niflheim

In Norse mythology, Niflheim is the afterworld inhabited by those who die of sickness and old age.

Acknowledgements

The author thanks the Canada Council for the Arts for support that helped provide time to work on this manuscript.

Professor Alfred Scharff Goldhaber and Judith Goldhaber were very helpful in providing personal background about Gerson Goldhaber.

Heartfelt thanks also to all the publishing team at the University of Alberta Press, who have been wonderful to work with once again.

The author also thanks the editors of the following publications in which some of these poems have appeared (sometimes in slightly different versions):

Literary magazines

CV2: "Love in three dimensions."

The New Quarterly: "Muscle of difficulty"; "Catechism"; "To the generations that will live a thousand years."

Arc: "The God of teapots"; "Persephone and I are underground."

Prairie Fire: "Last scattering surface"; "How to tell a Martian my heart is on the left"; "Ordinary matter."

Fiddlehead: "Tortoise and fern"; "Life adapts to inhospitable environments."

Antigonish Review: "Rectangularization of the morbidity curve"; "The movers' dilemma."

Freefall: "Allegory of Avarice"; "Allegory of Envy"; "Allegory of Sloth."

The New Quarterly: "Address"; "The end of greatness"; "Then death returns"; "$d = (X-x)^2 + (Y-y)^2 + (Z-z)^2 - c(T-t)^2$."

The Literary Review of Canada: "The helium thoughts."

Anthologies

40 below: "Optical molasses."

Bridges 2013 Poetry Anthology (Mathematical Poetry): "Zeno's paradox."

The Poet's Quest for God: "The god of sparrows."

Poems for an Anniversary (St. Thomas Press chapbook): "Day's eye."

E-zines

Truck: "Advice to the lovelorn."

London Grip: "The jeweller god"; "The muse of universes."

Poetry Daily: "Muscle of difficulty."

"The barber's paradox," "Zeno's paradox," and "Sand reckonings: Eubulides' paradox" also appeared in the paper "Barbers & Big Ideas: Paradox in Math and Poetry," by Alice Major in *The Journal of Mathematics and the Arts* 8.1–2, 2014.